◁◁ The transit of the SS *Ancon* on August 15, 1914, marked the official opening of the Panama Canal. ◁ Digging Gaillard Cut was punishingly hard work. Climbing the 154 stairs from the village of Culebra to the west bank of the Cut was the least of it. Temperatures at the bottom could reach 120°F. △ Landslides at the Cut were common. Months of work could be wiped out and tons of equipment destroyed overnight.

△ The average ship size at the Canal has grown tremendously since its early days. Four destroyers could fit in a single chamber at Pedro Miguel Locks in 1925. ▷ Heat and humidity notwithstanding, officials at the Canal maintained proper decorum at all times. Col. David D. Gaillard, who oversaw the excavation of the Cut that was later named for him, found time to have tea with his wife, Katherine, at their home in Culebra.

Portrait of the Panama Canal

FROM CONSTRUCTION TO THE TWENTY-FIRST CENTURY

~ *Centennial Edition* ~

Text by WILLIAM FRIAR

~

Foreword by GEORGE R. GOETHALS

GRAPHIC ARTS CENTER PUBLISHING ®

ISBN-10: 1-55868-746-7 ISBN-13: 978-1-55868-746-2
Library of Congress Catalog Number 99-63186
Published by Graphic Arts Center Publishing®
    An imprint of Graphic Arts Center Publishing Company
    P.O. Box 10306 • Portland, Oregon 97296-0306
    503/226-2402 • www.gacpc.com
President • Charles M. Hopkins
Associate Publisher • Douglas A. Pfeiffer
Editorial Staff • Timothy W. Frew, Tricia Brown, Jean Andrews,
    Kathy Howard, Jean Bond-Slaughter
Designer • Robert Reynolds
Production Staff • Richard L. Owsiany, Joanna Goebel
Book Manufacturing • Lincoln & Allen Company
Printed and Bound in the United States of America
Third edition—Celebrating the Centennial of the Panama Canal
Fourth Printing

~

Text © MCMXCVI and MMIII by William Friar
Foreword © MCMXCIX by George R. Goethals
Canal Map page 12-13 & Map of Panama page 13
    by Eureka Cartography
World Map page 7 & Profile Map page 12*
Photographs pages 5 & 23 from Panama Canal archives*
Photograph pages 22 & 30 © Panama Stock Photo/Fran Casey Jr.
Photographs pages 18, 32, & 71 © MCMXCVI by Maxine Cass
Photographs pages 19, 64, & 79, and Back Cover photograph
    by Jaime Fernandez*
Photograph page 9 © MMIII by William Friar
Front Cover photograph and photographs pages 16, 17, 27, 28,
    30, 34, 35, 36–37, 39, 40–41, 42, 50, 59, 62, 63, 65, 66,
    & 75 by Don Goode*
Photographs pages 44–45, 46, & 56–57
    by Armando de Gracia*
Photographs pages i, ii, iii, iv, 1, 6, 20–21, 25, 26, & 31
    by Ernest "Red" Hallen*
Photograph page 51 © La Prense/Tito Herrara
Photographs pages 67 & 78 © MCMXCVI by Dave G. Houser
Photographs pages 11, 24, 29, 33, 38, 43*, 48, 49, 68–69, 70,
    & 74 © by Kevin Jenkins
Photographs pages 2, 8, 43, 47, 52–53, 58, 60–62, 76–77, &
    80 by Kevin Jenkins*
Photograph page 15 by Melvin Kennedy*
Photographs pages 48 & 49 © MCMXCVI by Harvey Lloyd
Photographs pages 10, 54, & 55
    © MCMXCVI by Vanessa A. Puniak
Photograph pages 4 © MCMXCVI by James C. Simmons
Photograph page 14 photographer unknown*
Photograph pages 72–73 by Jaime Yau,
    Courtesy of the Panama Canal Authority
*Courtesy of the Panama Canal Commission

◁ ◁ *A PANAMAX ship, the largest type of vessel normally transiting the Canal, moves through Gaillard Cut. The Cut was widened in 2001 so these huge ships could pass when traveling in opposite directions.* △ *Sandy beaches beckon.*

Foreword

A 1913 edition of the British magazine *Puck* features a two-page, full-color cartoon depicting Uncle Sam astride the Isthmus of Panama and the nearly completed Panama Canal. The cartoon also shows the pyramids of Egypt, the hanging gardens of Babylon, and other familiar human creations. Its caption reads, "The seven wonders of the world salute the eighty." Cradled in Uncle Sam's arm is a figure, a saluting, white-haired, white-suited mustachioed man with the name "Goethals" printed on his collar. This is my great-grandfather, chief engineer of the Panama Canal.

Though my proud reaction to this cartoon is nepotistic, and not everyone shares the cartoon's unabashed jingoistic sentiments, few would quarrel with hailing the Panama Canal as a wonder of the world.

Built at the beginning of the twentieth century, the Canal has become a vital link in the global economy of the twenty-first. The environmental and social disruptions caused by its construction long behind us, the Canal exists in harmony with the fragile rain forest ecosystem of the Isthmus. While there were many changes in the Canal and its operations through the years, much of the Canal uses original equipment. The locks, the gates, the dams, and the breakwaters were built to last by very smart people.

While visiting and transiting the Canal are the only ways of absorbing its full majesty, beauty, and significance, William Friar's remarkable *Portrait of the Panama Canal* provides an unusually thorough and faithful perspective on the Canal as it paints a compelling portrait of the skill and sacrifice of those who built and have operated this crucial waterway.

This is a personal book as well as an account of an engineering and economic marvel. William Friar makes clear his own attachment to the former Canal Zone and the Canal itself. My attachments are personal as well. On my first trip to Panama in 1996, William Friar's mother, Willie K. Friar, the now-retired director of public affairs for the Panama Canal Commission, encouraged me to explore the workings of the Canal in great detail. I saw the lock gate operations through the eyes of the men and women running them every day. I am grateful to Ms. Friar for helping me make a connection to an important piece of family history.

Upon seeing the Canal, I was reminded of the first time I understood that my great-grandfather had accomplished something significant in a faraway place. In Vineyard Haven, Massachusetts, where I grew up, directly across from the elementary school stands the General George W. Goethals American Legion Post. Early in my kindergarten year, a friend led me across the street to climb the trees in front of the legion hall. Later, a teacher scolded us, saying the trees belonged to George W. Goethals and that we had no business on them. When he asked me my name and I said George Goethals, the roomful of students howled with laughter and the teacher looked most displeased. That is all I remember from this frightening experience. But it did lead me to learn more about my famous forebear.

△ *The Panama Canal actually opened six months ahead of schedule and under budget. The credit for this belongs to George W. Goethals, who was the chief engineer of the Canal during the last seven years of its construction.*

For decades my curiosity about the Canal went unsatisfied. When I finally experienced the Canal, I was awestruck by its genius and its significance. William Friar's *Portrait of the Panama Canal,* revised to recognize the historic transfer of Canal operations to the Republic of Panama, offers an enticing preview of the Canal for those who plan to visit Panama, and a souvenir for those who have experienced it. I am delighted to welcome you to Friar's *Portrait* and to share my enthusiasm for this marvelous achievement of the human spirit.

—George R. Goethals,
Williamstown, Massachusetts, May 12, 1999

The Canal Zone

We live in Paradise
and our handmaiden is Felicity.
—AN ANONYMOUS CANAL ZONE RESIDENT,
circa 1914, when the Canal was finished

It was strange growing up in the Canal Zone, so strange that my earliest memory of it actually pre-dates my arrival there. I was six years old, and my mother, brother, and I were en route from New Orleans to the Canal Zone on the *Cristobal,* one of three ships owned by the Panama Canal Company, my mother's new employer. The year was 1969.

It must have been a night or two before we docked at the *Cristobal'*s namesake port, on the Atlantic side of the Isthmus of Panama. We were still sailing far out at sea, and I was asleep on the top bunk of our cabin. All was quiet except for the slow splash of the waves breaking against the hull and the drone of the rotary fan swinging sleepily back and forth on a ledge. Apparently it wasn't helping much, because while I slept heavy air poured into the cabin, layer upon layer, until the weight of it squeezed me awake. I had never felt anything like it before. The air was hot, wet, suffocating, as if I were wrapped in soggy towels in a steam-filled sauna.

Years later, whenever I flew home from the States on college break, that familiar blast of humid air would be the first thing to greet me when I got off the plane. It always felt like an embrace.

I lived on the Pacific side of the Isthmus—first in Diablo, then Balboa, then Balboa Heights—until I left for the States to attend college at age eighteen. People always ask me what it was like to live in the Canal Zone. There is really no good way to explain it. Just the sheer physical facts of the place are baffling, thanks to the snake-like shape of the Isthmus. Imagine a land where the Atlantic Ocean is north of the Pacific Ocean, and where you can take a swim in both, if you want, on the same day. Where a ship transiting the Canal from the Atlantic to the Pacific will end up east of where it started. Where you can watch the sun rise over the Pacific.

Life in the Canal Zone was equally peculiar. This strip of land fifty miles long and ten miles wide was created in 1903 as the operating area for Americans building their Canal, which was to run down the middle of it. Even before the Canal was finished in 1914, the Zone took on an identity. It became a community, or, rather, a cluster of communities. As time went by, just living on different sides of the Isthmus placed you in different worlds. Those who lived near the Pacific talked about going to "the Other Side," and vice versa, though the distance between these alien worlds was less than fifty miles.

Canal Zone residents lived under United States law. We had our own police force, court system, even stamps. The American kids who grew up there—and most of the Zonians, as we were called, were American—looked and sounded like any kids growing up in the suburbs anywhere in the United States; yet there was something different about us. To this day some of my old Zonian friends and I do

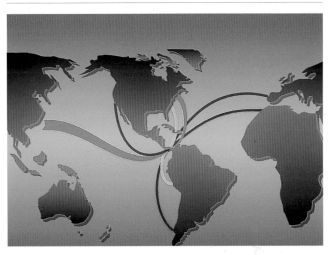

◁ *Massive as pyramids and intricate as the workings of a fine watch, the Canal locks took four years to build. The lock walls are one thousand feet long and taller than a six-story building.* △ *Panama is called "the crossroads of the world."*

not feel completely at home in the United States, or any other place, for that matter.

Canal Zone kids joined the Boy Scouts or Little League—and once a year paddled the length of the Canal in Native dugout canoes called *cayucos.* We were taught in English at American schools, but every day we spoke Spanish. Some of us were bilingual, and all of us could at least get by with a sloppy but fluent version of "Spanglish." We spiced our teenage conversation with Panamanian slang and curse words, as well as a smattering of *Bajun,* a Barbadian dialect we learned from the descendants of West Indian construction-era laborers.

About forty-six thousand people lived in the Canal Zone during its heyday. Thirty thousand were American military and their dependents, who lived apart from the Zonians on military bases and generally stayed on the Isthmus only a few years. The remaining sixteen thousand were employees of the Panama Canal Company, the United States government agency that ran the Canal, and their families.

Six thousand of these were of West Indian descent, mostly Panamanian citizens. The rest were Americans, about thirty-five hundred Canal employees and sixty-five hundred dependents. These were the ones most people think of as Zonians. Some even had parents and grandparents who had worked on or helped build the Canal.

△ *The annual Cayuco Race is a three-day event in which contestants paddle streamlined versions of dugout canoes through the Canal. Paddlers hang on to ropes for safety in lock chambers, which fill and empty in about ten minutes.*

Some red-blooded American Zonians, who went religiously to the high school football game on Friday and church on Sunday, might take exception to a description of the Canal Zone as a socialist paradise, but in many ways that is what it was. Since everyone in the Zone had to be an employee of the Company, the Canal Zone had no unemployment, no poverty, and almost no crime. Everyone worked for the government, and no one owned the house he or she lived in. Class distinctions were much less pronounced in the Zone than they are in the States. Housing assignments were based solely on seniority, so a lawyer and an electrician could quite conceivably find themselves living across the hall from each other in one of the Company's modest duplexes.

About eighty-seven hundred Panamanians commuted to the Zone for work. Most came from Panama City, which, with its glittering high-rises, garish billboards, and swank clothing stores—not to mention striking contrasts between the extremely rich and the dirt poor—seemed closer to an American city than any other town in the Zone did. In high school, my friends and I liked to hop in someone's family car at lunchtime and head "Downtown," as everyone referred to Panama City, to McDonald's. That all-American franchise was not allowed in the American Canal Zone.

It was a great place to be a kid. There were hundreds of miles of near-deserted beaches to explore along the coast, with bathtub-warm waters you could swim in 365 days a year. If you were lucky, you knew someone who owned a boat, and you went waterskiing on a huge lake made by the Canal builders. As you sliced through freshwater over what had once been mountainous jungle, you shared the lake with distant ships silently gliding past each other as they made their way through the Canal, some heading for the Atlantic, others for the Pacific.

The Canal Zone, like any community, had its limitations. It was clean and orderly to the point of being sterile. Like any small-town world, it could be claustrophobic and cliquish. Some, including myself, never felt fully a part of that world.

Still, the Canal Zone was my home, and now it is gone. But the Panama Canal itself, the incredible creation for which every Company employee ultimately worked, is still there. This is its story.

The Dream of a Canal

> *. . . Silent, upon a peak in Darien.*
> —JOHN KEATS,
> "On First Looking into Chapman's Homer"

At ten o'clock in the morning on September 26, 1513, a Spanish explorer named Vasco Núñez de Balboa climbed one last hill, stepped out of the dense Darien jungle, and looked at an immense expanse of blue water never before seen by European eyes.

What Balboa had "discovered" was the Pacific, separated from the Atlantic by an isthmus of nearly impenetrable, mountainous jungle. In places, the sliver of green separating the two vast bodies of blue was less than fifty miles thick. If that narrow stretch of land could be pierced, the Atlantic and Pacific would be linked, opening up new worlds to explore.

In 1519, the Spanish governor of the region, Pedro Arias de Avila, better known as Pedrarias, founded the city of Panama. The name comes from a Cueva Indian word that means "the place where many fish are taken." Pedrarias has gone down in history as a sinister, treacherous figure. To forestall a threat to his power, Pedrarias accused Balboa of treason and had him beheaded in the land of his greatest triumph.

Sixteenth-century Spaniards actually toyed with the idea of a transoceanic canal through Central America, but no attempt was made or would even have been possible, given the technology of the time. Panama for the Spanish became a different kind of thoroughfare between the oceans, as the *camino real,* or royal road—in reality a muddy mule trail—for transporting conquistadores' spoils from the rape of the New World. Pearls, the gold of the Incas, and the silver of the Andes were all shipped across the Isthmus to galleons waiting on the Atlantic side.

Still, one can trace the dream of a canal between the oceans to Balboa's first glimpse of the Pacific. But it would take more than four hundred years for that dream to become a reality. Doing so required an effort unprecedented in human history. The price alone was staggering: hundreds of millions of dollars, tens of thousands of lives, and immeasurable toil. But when it finally opened on August 15, 1914, the Panama Canal took its place among the most magnificent engineering works of all time.

Gold Rush

I have no time to give reasons, but . . . for no consideration come this route. I have nothing to say for the other routes but do not take this one.
—A MASSACHUSETTS MAN,
writing home after crossing the Isthmus of Panama

Panama has a habit of slumbering for ages, only to be jolted awake by the outside world: first by the Spanish in the early sixteenth century; then by the English privateer Sir Francis Drake, who raided the Spanish forts in 1572 and 1595, dying on the Isthmus that same year; then by the Welsh pirate Henry Morgan, who sacked the old Spanish city of Panama in 1671.

For two centuries after that, the world pretty much ignored Panama. Then in 1848, an American carpenter in California's Sierra Nevada found gold in a millstream. When the news reached the East Coast of the United States, the promise of riches once again forced Panama to wake up from its tropical torpor. And once again the focus of activity was finding a way across the Isthmus.

Those who became known as the forty-niners had three ways to reach California: across the

△ *Cana, deep in the Darien jungle, was a rich gold mine for hundreds of years. All that's left now is rusting machinery. It has become a prime ecotourist spot, considered one of the world's top ten birding destinations.*

still-wild heart of the North American continent, which would not be tamed by the railroad for another twenty years; around the tip of South America; or across the Central American Isthmus.

The most popular Central American route was a grueling journey by mule, dugout canoe, and foot across Panama. Taking this route meant surviving a fifty-mile gauntlet though the boiling heat of a nearly impenetrable rain forest lousy with disease, wild animals, and merciless insects. But if you survived, you shaved precious time and thousands of miles off your journey to the goldfields. Traveling from New York to San Francisco by way of Cape

Horn meant a voyage of thirteen thousand miles. By detouring through Panama, you could cut the trip to five thousand miles. For twenty-seven thousand lucky Forty-niners, the risk was worth it.

The steamer *Falcon* unloaded the first group of two hundred would-be prospectors at the mouth of Panama's Chagres River on January 7, 1849, much to the astonishment of the locals. Miraculously enough, every one of these ill-prepared adventurers survived the fifty-mile trek across the Isthmus. Untold others would succumb to one or another of the tropical diseases—malaria, dysentery, cholera, yellow fever—that over the centuries had given Panama a well-deserved reputation as a death trap.

In 1848, before the gold rush, a prescient American merchant by the name of William Henry

△ *In 1513, Vasco Núñez de Balboa became the first European to see the Pacific Ocean. He claimed the* Mar del Sur *(Sea of the South) and all lands touching it for Spain. In Panama City, a statue in his honor gazes out at the Pacific.*

Aspinwall negotiated with the government of New Granada, as Colombia was then known, to buy the rights to build a railroad across the Isthmus of Panama, one of its "departments" or provinces.

Plans called for a one-track railroad stretching across the Isthmus for forty-seven and a half miles. By the time work was completed in 1855, builders had constructed 304 bridges and culverts along the railroad's hard won length. If the Panama Canal had never been built, the railroad alone would have assured the Isthmus a place in history as the site of one of the world's great engineering feats. That railroad proved to be one of the most expensive per

mile in history, finally costing eight million dollars to build, a huge sum in those days.

The railroad was also costly in human lives. It has often been claimed that every tie along its length represented one dead worker. There is no truth to that colorful saying, but no one will ever know how many thousands died building the line, since the Panama Railroad Company kept no records of the death count. When the railroad was finally completed in 1855, the Mount Hope railroad cemetery held six thousand graves. Thousands more bodies were disposed of elsewhere. So many died that the Panama Railroad Company started a booming side-business pickling cadavers in barrels and shipping them to medical schools and hospitals around the world.

Most workers were Irish and Chinese. The Irish died in droves; the Chinese did not fare much better. Besides succumbing to disease, dozens of them committed suicide, though the reason is subject to debate. "Melancholia" is common after a bout of malaria, which was endemic. But many of the Chinese laborers came to the Isthmus addicted to opium, and the scandalized company cut off their supply, plunging addicts into withdrawal. Whatever the cause, desperate Chinese workers killed themselves by hanging, drowning, or impaling themselves on bamboo poles.

Then there was the work itself. It was not just a matter of building a railroad through thick, dangerous jungle; numerous swamps and rivers lay in its path. Most infamous of these was the legendary Black Swamp on the Atlantic side of the Isthmus, reputed to be bottomless. It wasn't: surveyors finally struck bottom at 185 feet. But the swamp's appetite certainly seemed bottomless, consuming ton after ton of rocks and tree trunks as frustrated engineers tried to build a foundation for the railroad. Finally, flatcars were chained together and submerged, to create a platform upon which rocks were heaped, creating a floating roadbed.

The greatest obstacle was the Chagres, mightiest of the rivers of the Isthmus and hardest to contend with. In a tropical downpour, the Chagres could rise ten feet in an hour, turning it from a modest stream to a raging torrent before one's eyes. The railroad had to cross this river at a point where it was three hundred feet wide. As railroad builders learned to their chagrin when a flood washed away their first bridge, the river here could rise forty feet overnight. Finally, a sturdy iron bridge 625 feet long was constructed high above the river, but even this was occasionally damaged by severe floods.

In July 1852, with the railroad only half finished, a cholera epidemic broke out, spreading so quickly that many laborers were struck down while working. The lucky ones managed to drag themselves up the tracks, where they were found and taken to the local hospital. Others collapsed and died on the spot, to be eaten by ants and land crabs. All but two of the fifty American engineers and technical staff died.

At the height of the epidemic, several hundred soldiers from the United States Army's Fourth Infantry arrived on the Isthmus with their wives and children, on the way to duty in California. The group made it to the end of the partially completed rail line, halfway across the Isthmus at Barbacoas, but by then several people had come down with cholera. They decided to split the group in two, with the healthy soldiers enduring a three-day march from Gorgona to Panama City. Women, children, and the sick were to travel by mule. But the local mule-owner who had promised to transport them for eleven cents a pound found that others were willing to pay almost twice as much to rent his mules. He reneged on the deal. The detachment was stranded at the village of Las Cruces for five days, during which a dozen men died. The captain in charge of the group finally struck a new deal with the mule-owner, and at last the detachment completed its journey along the rocky, muddy trail. In all, 150 men, women, and children died during or after the crossing.

"The horrors of the road in the rainy season are beyond description," wrote the captain in charge of the detachment, who would never forget this nightmarish trek, reportedly more affected by it than any of his many future battles. His name was Ulysses S. Grant. Later, as president of the United States, he would commission the first serious surveys for a possible interoceanic canal. Explorers made seven expeditions to determine the best route across Central America between 1870 and 1875. In 1875, Grant's Inter-oceanic Canal Commission issued its recommendations. Its verdict was unanimous: Nicaragua.

The French

…and I maintain that Panama will be easier to make, easier to complete, and easier to keep up than Suez.
—Ferdinand de Lesseps

Even with a route chosen, the United States had no definite plans to build a canal and was not even sure how it could be done. But without ever having set foot in the Americas, the solution was perfectly obvious to one man: Vicomte Ferdinand de Lesseps.

By the time he turned his attention to the question of a Central American canal, de Lesseps was already a legend. He was responsible for one of the most awe-inspiring engineering feats of his age: building

△ *The Administrator's House, home to the Canal's chief executive, was the chief engineer's residence during construction days. It was dismantled in 1914 and moved from its location above the Gaillard Cut to Balboa Heights.*

a sea-level canal across the Isthmus of Suez to link the Mediterranean and Red Seas. That he did so was almost miraculous. De Lesseps had been a diplomat from the age of nineteen, and had no engineering background, no experience directing public works, nothing that would seem to qualify him for this incredible task.

Nothing, that is, except for his charisma, his air of authority, and the respect and even love he won from all who came in contact with him. The canal was built almost by the sheer force of his will.

The canal took fifteen years to complete, opening just two days before de Lesseps' sixty-fourth

Because the Panama Canal links the Atlantic Ocean (to the east) with the Pacific Ocean (to the west), it is perceived as running east-west. However, the land angles so that the Canal actually runs northwest to southeast. This makes it especially confusing because the Canal runs neither true north-south nor true east-west.

Started by the French, who attempted to dig a sea-level canal, the Panama Canal was taken over by the United States government after the French company went bankrupt. The Americans changed the concept to a lock canal. Only about one-half of the excavation accomplished by the French was usable by the Americans.

The Panama Canal could accurately be described as a water elevator. When ships enter the locks, they are raised in several steps eighty-five feet above sea level, then are lowered back down to sea level by the locks at the other end of the Canal to continue their journey. The transit from ocean to ocean takes from eight to ten hours.

Amazingly, the water enters and leaves the locks by the force of gravity alone; no pumps are used. All the water for each passage is lost to the ocean after each ship passes through.

Considered one of the world's greatest engineering feats of all time, the Panama Canal was opened to ocean-going vessels in 1914.

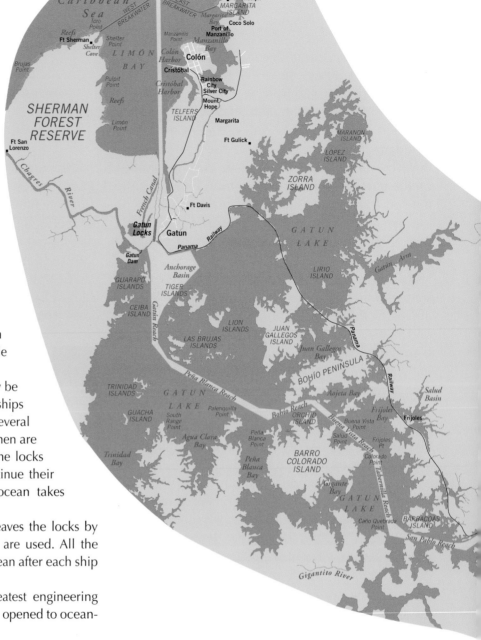

PANAMA CANAL PROFILE

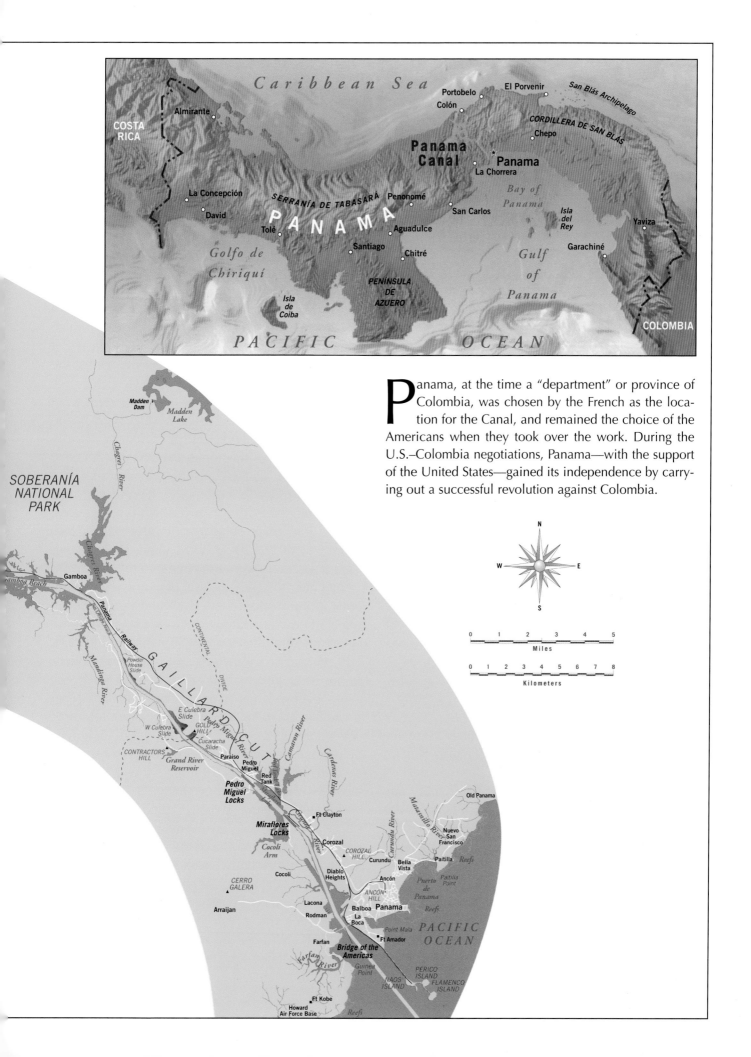

Caribbean Sea

COSTA
RICA

Almirante

Portobelo   El Porvenir   San Blás Archipelago

Colón

CORDILLERA DE SAN BLÁS

Chepo

**Panama
Canal**   ★ **Panama**

La Chorrera

La Concepción   *SERRANÍA DE TABASARÁ*   Penonomé

David   *Bay of
Panama*

P A N A M A   San Carlos   *Isla
del
Rey*   Yaviza

Tolé   Aguadulce

*Golfo de
Chiriquí*   Santiago   Chitré   *Gulf
of
Panama*   Garachiné

*Isla
de
Coiba*   **PENÍNSULA
DE
AZUERO**   COLOMBIA

P A C I F I C   *O C E A N*

anama, at the time a "department" or province of
Colombia, was chosen by the French as the location for the Canal, and remained the choice of the
Americans when they took over the work. During the
U.S.–Colombia negotiations, Panama—with the support
of the United States—gained its independence by carry-
ing out a successful revolution against Colombia.

birthday, on November 17, 1869. From then on, he was known as "Le Grand Français"—The Great Frenchman—by his adoring countrymen.

Hungry for another triumph, he insisted he could build a similar sea-level canal at Panama. Ignoring objections of thoughtful engineers who said this was impossible, or at least a prohibitively difficult, expensive undertaking, he relied on flawed surveys by his own associates and on his supreme self-confidence.

De Lesseps parted the sands of Suez with his charm, but the jungles of Panama were a different proposition. At Suez, the terrain was flat, sandy, and no higher than fifty feet above sea level. Panama is mountainous: The maximum elevation at the site of the planned French canal was three hundred feet. Digging a canal through Panama meant breaking

△ *The only person to swim the entire Canal, adventurer Richard Halliburton did it over three days in 1928, paying a record-low toll of 36 cents. A sharpshooter, watching for caimans and crocodiles, followed in a rowboat.*

through the mountains of the Continental Divide, the rock fortress wall that splits the American continents down the middle.

As any railroad worker could have told him, digging a ditch at the Isthmus meant contending with solid rock, swamps, clay that stuck like glue, and earth that could quickly turn into an avalanche of mud, burying months of work in a matter of hours. Even when it did not cause landslides, the rain was overwhelming: the wet season in Panama lasts eight months, dumping as much as twelve *feet*—not inches—a year. The humidity approaches 100 percent in the rainy season, air so liquid that just sitting

still in the shade can be suffocating, let alone doing backbreaking work under the hot sun.

And then there was the Chagres. A canal across the Isthmus had to cross this seemingly untamable river. The prospect was so daunting that one proposal called for building a stone viaduct nineteen hundred feet long above and across the river, so ships could sail over the Chagres on a kind of water bridge. The other option was to find a way to rein the river in; that is what the French proposed to do. To accomplish this, they would have to build one of the largest earthen dams in history.

In 1882, the first year of excavation, a strong earthquake struck Panama, a rare occurrence there, which inflicted major damage on the Panama Railroad. It was as if nature itself were conspiring against the canal builders.

But small menaces were the most important ones. The workers at Panama, unlike those at Suez, had to share every waking and sleeping moment with tropical insects and wild animals. Panama was home to pumas, jaguars, and a whole menagerie of snakes—including the deadly bushmaster, fer-de-lance, and coral. The place also crawled with mosquitoes, spiders, ants, gnats, chiggers, and other pests. Canal laborers slept with their boots under their pillows to keep scorpions from crawling into them at night.

Long after the French canal plan and many thousands of the workers were dead and buried, scientists learned that mosquitoes were by far the most dangerous of all the torments Panama had to offer. They transmitted Panama's two deadliest killers: yellow fever and malaria. During the French era, it was believed that malaria—or "bad air"—was spread by breathing poisoned air, especially noxious marsh gases blown by the wind to human settlements. Yellow fever was believed to be a kind of contamination spread through sewage and any contact with an infected person.

Although malaria actually killed more people, yellow fever was the more terrifying disease because of its horrific symptoms and the speed with which it dispatched its victims. Death within three days of the onset of symptoms was not uncommon.

A person infected with yellow fever would begin to shiver and run a high fever, develop unquenchable thirst, and suffer severe aches and pains. Within a day or two, the victim's face and eyes turned yellow.

14

In the final stages, the patient was wracked with *vómito negro,* vomiting up black blood. Add to this the occasional epidemic of such deadly scourges as cholera, beriberi, smallpox, and typhoid fever, and it was understandable why many who came to Panama regarded it hell on earth.

This is what awaited French laborers, who plunged into the work like soldiers into battle. And like soldiers they died. Accurate records were not kept, but estimates are that twenty thousand men, women, and children lost their lives between the arrival of the first French canal-builders in January 1881 and the project's collapse eight years later.

The French did all they could to prevent and treat the diseases, but without an understanding of how malaria and yellow fever were spread they were helpless. In fact, some of the things they did actually encouraged the spread of these scourges. The French hospital on Ancon Hill near Panama City, for instance, was actually a breeding ground for disease. In the hospital gardens, pottery rings filled with water were placed around plants to protect them from ants. And in the hospital, staff members placed containers of water under the legs of patients' beds to keep ants from pestering the sick. Mosquitoes laid their eggs in these containers; when they hatched, armies of mosquitoes bit the sick patients, picking up yellow fever and malaria parasites, and then flew off to infect others.

In spite of the obstacles, the French moved fifty million cubic meters of earth and rock—two-thirds the total excavation at Suez—in their quest for a sea-level canal. Though only a little over half this work would help the Americans who succeeded them, it was an impressive accomplishment. But the French plan for a sea-level canal was fatally flawed from the start, through a combination of poor planning and absurdly low estimates of its cost, much of it attributable to the overconfidence, even arrogance, of de Lesseps.

Once the work began, the French compounded their mistakes and misjudgments. Equipment and machinery shipped to Panama was of a bewildering variety. Railroad track laid by the French, for instance, was of a different gauge from that of the Panama Railroad. Work was farmed out to subcontractors—so many that one hill on the route is still called Contractors Hill. Some contractors were completely unqualified for the work.

The most difficult excavation was in a gap in the Continental Divide called Culebra, meaning "snake." The terrain at Culebra, layers of clay alternating with slabs of rock, meant that when the rains came, large chunks of the mountain crashed down into the ditch. Sometimes the cascading earth took track, railcars, and steam shovels with it. Months of work could be wiped out with a single rainstorm. The only way to prevent slides was to dig the Cut at sharper and sharper angles, in search of what engineers call the "angle of repose." The V-shaped excavation became wider and wider at the top; each redrawing of the plans meant a massive increase in the total amount of spoil that had to be removed. The French stumbled badly at Culebra, never even figuring out how to dispose of the dirt and rock they did manage to remove.

△ *The macaw is but one example of Panama's spectacular bird life. Jungles and highlands offer many surprises, including the golden frogs and square trees that inhabit El Valle, a cool, fertile region in the crater of an extinct volcano.*

Finally, the money ran out. The whole enterprise collapsed on February 4, 1889, sending up a dark cloud of scandal and bankruptcy that lingered for years over the French republic. Hundreds of thousands of ordinary French citizens who had invested in the enterprise lost everything. An investigation revealed that government officials and journalists had been bribed to buy support of the undertaking.

The scandal drove the French government from power. Careers and reputations were destroyed, including that of Le Grand Français himself. When he died on December 7, 1894, only his family and a few friends attended his burial.

The final cost of the failed attempt to build a sea-level canal across the Isthmus has been reckoned at 287 million dollars, up to that time the most expensive peacetime project in history. Only a tenth of the necessary excavation had been completed.

The Dream Lives

> *FATE NEWS BAD POWERFUL TIGER.*
> *URGE VAPOR COLÓN.*
> —FUTURE PRESIDENT MANUEL AMADOR GUERRERO
> Coded message requesting an American gunboat
> to aid in the revolution against Colombia.

France's self-confidence was deeply shaken by this debacle and the even more momentous Dreyfus

△ *Northbound: A container ship leaves Miraflores Locks on its way to Pedro Miguel Locks, on the opposite shore of Miraflores Lake, a man-made freshwater lake one mile wide. The ship is headed for the Atlantic Ocean.*

spying scandal that followed. Meanwhile, a young United States was brimming with enthusiasm and self-confidence, looking for ways to make its presence more strongly felt on the world scene. For some fifty years, the United States had toyed with the idea of a canal across Central America. With the collapse of the French effort at Panama, it took up the matter in earnest.

Then, in September of 1901, the assassination of William McKinley brought Vice President Theodore Roosevelt to the White House. For more than a decade, Roosevelt had been one of the strongest advocates for a canal connecting the Atlantic and

Pacific Oceans. To Roosevelt, the canal was the key to making the United States a major sea power, with a strong naval presence on both oceans. Its importance to world commerce ran a distant second in his thinking.

In late 1901, the drive for a United States–built Isthmian canal was unstoppable. The question was where it should be built. Since the surveys of the 1870s, public sentiment had run strongly in favor of Nicaragua, even though engineering data suggested that a Panama canal would be less expensive and take less time to build than one at Nicaragua. But Nicaragua's climate was more hospitable and less disease ridden than Panama's, and the route did not carry the stench of corruption and disaster that tainted the French effort at Panama.

In November, the results of a presidential study commissioned by McKinley two years earlier were released. The study came out in favor of the Nicaragua route, but only out of concern for how much it would cost to buy the French holdings in Panama, valued at 109 million dollars. If technical matters were the only consideration, the commission reported, Panama was the better route.

Enter Philippe Bunau-Varilla, the former director general of the Compagnie Universelle du Canal Interocéanique, who became the primary lobbyist for the Panama route. Bunau-Varilla, a major stockholder in the failed French effort, stood to lose millions of francs if there were no bail-out of the bankrupt company. But monetary considerations aside, for Bunau-Varilla the idea of a canal across Panama had become an almost holy crusade. He was determined to convince the Americans to take up the cause.

A battle over the two routes raged in Congress, the press, and the nation's clubrooms for months. It seemed inevitable that the Nicaragua faction would win eventually, in spite of Bunau-Varilla's extraordinary efforts. He convinced the Compagnie Nouvelle, the successor to the bankrupt French company, to offer its assets to the Americans for forty million dollars, less than 40 percent of their estimated value. He toured the country, giving speeches to all who would hear him. He dined with engineers and shipowners. He smoked cigars with senators. He charmed bank presidents and chambers of commerce. He mailed out pamphlets.

Bunau-Varilla pounded away at the fact that Nicaragua had fourteen volcanoes, some of them active. Volcanoes were depicted in Nicaragua's coat of arms and postage stamps, a clear sign of their central place in the country's history, he declared. He argued that an eruption could destroy a Nicaraguan canal.

Then on May 8, 1902, nature responded to its cue, as it so often seemed to do in matters concerning Panama. The eruption of Mount Pelée, a volcano on the Caribbean island of Martinique, wiped out a city of nearly thirty thousand people in a matter of minutes. Martinique is fifteen hundred miles from Nicaragua, which may have seemed a world away to some. But on May 14, Momotombo, a volcano believed to be extinct, erupted in Nicaragua itself. On May 20, Mount Pelée erupted again. Then yet another volcano erupted, on the nearby island of St. Vincent.

Debate on the canal route was scheduled to start June 4, 1902. Still, the votes seemed to be against Panama. The Nicaraguan embassy insisted, untruthfully, that there had been no eruption in Nicaragua. Senate advocates of the Nicaragua route questioned the motives of Bunau-Varilla and his colleagues. They described Panama as both politically and seismically unstable, the latter based on the earthquake during the first year of French excavation, in 1882. In spite of Bunau-Varilla's efforts, it seemed the dream of a Panama Canal would die in Washington.

Then, just a week before the Senate vote, Bunau-Varilla seized on one last, desperate gambit. He remembered a one-centavo Nicaraguan stamp that showed an erupting volcano in the background. It was Momotombo, the volcano that had erupted again on May 14. He combed the stamp shops of Washington, managing to find ninety of the incriminating bits of paper, one for each senator. He pasted each stamp to a sheet of paper and mailed them to the Senate offices, where they arrived on June 16, just three days before the vote.

No one knows whether this last bit of propaganda tipped the balance, but on the afternoon of the nineteenth, the Senate voted forty-two to thirty-four in favor of the Panama route. Five votes less, and the Isthmian canal might have been dug through Nicaragua. That narrow victory was confirmed resoundingly in the House on June 26, by a vote of

259 to 8. President Roosevelt signed the bill into law two days later, on June 28, 1902. The Panama Canal was going to become a reality after all.

Or so it seemed. There was one remaining obstacle: arranging a treaty with the Colombian government, which still claimed Panama as one of its "departments," or provinces.

Though eager to reach an agreement with the United States, the Colombian government kept withdrawing its negotiators. The United States ended up working with three diplomats in all, each with his own agenda and personality. And communications with Colombia, for both diplomats and U. S. officials, were extremely difficult and time-consuming, as turn-of-the-century Bogota was one of the most remote and inaccessible cities in the world, set on an inland

△ Southbound: A container ship exits Miraflores Locks and enters the Pacific. Panama's mountains make the locks necessary, not the oceans. There is no difference in the "height" of the two oceans: sea level is sea level.

plateau eighty-six hundred feet high. Reaching it from the coast meant weeks of hard traveling. Letters between Bogota and Washington took months to arrive.

The Colombian government was also unstable, wracked by political infighting and a chronic civil war—not the best conditions for working out a treaty with major implications for both countries.

There were several sticking points in the substance of the treaty. Colombia was worried that the United States might not respect Colombia's sovereignty over the land through which the canal would pass. As if to confirm that country's fears, Roosevelt actually sent U.S. Marines to Panama during the negotiations, without Colombia's permission, to protect the Panama Railroad during a brief uprising against the Colombian government.

Another concern was that Colombia had been excluded from the agreement the United States had worked out with the French canal company. Colombia had a right to part of the $40-million settlement, since the French company's agreement with Colombia explicitly forbade it from selling its right to build the canal to any foreign power. But Colombia was not offered a single dime of the settlement, the most expensive land transaction in United States history up to that point.

By all accounts, the Colombian government was extremely hard to negotiate with. But the United

△ *Bounded on two sides by oceans and with a man-made waterway running through its heart, water is as important to Panama's identity as land is. Vessels ranging from great ocean liners to dugout canoes use the waterways.*

States government treated Colombia, about which it knew little, with contempt. Roosevelt called Colombian officials "bandits" and "jack rabbits."

It was out of this atmosphere that the Republic of Panama was born. Exactly how and why this happened was a subject of great controversy and debate at the time, and it can still provoke arguments to this day. The most common impression of what happened is that Roosevelt, frustrated by dealing with the Colombians, seized Panama in a colonialist display of "gunboat diplomacy." Roosevelt himself encouraged this interpretation, most notoriously in a 1911 speech in which he boasted: "I took the

Isthmus, started the canal and then left Congress not to debate the canal, but to debate me."

This declaration was both arrogant and far from the whole truth, though that truth will probably never be fully known. The story that historians have pieced together is subtle and complex, involving many parties with many different motivations. In a nutshell it is this:

Panama had always been a neglected, backwater department of Colombia, separated from the rest of the country by the dense Darien jungle. Panama was infamous for its political turmoil, with fifty-three uprisings in the fifty-seven years before it finally achieved its independence in 1903.

In July 1903, a small group of Panama Railroad employees, including a few Americans, met in secret to plan a revolution. Because of the railroad connection, later observers speculated that the impetus for the uprising came from William Nelson Cromwell, an American lawyer who was the French canal company's chief lobbyist in the United States, as well as general counsel for the Panama Railroad. There is no way to confirm exactly what his role was, but it is known that the conspirators were in contact with Cromwell, and relied on him for help in gaining United States support for their plot.

Ultimately, he let them down, apparently fearful that Colombia might have learned of their plans, which put at risk all of the French holdings in Panama. Manuel Amador Guerrero, the elderly physician for the railroad, was sent as an emissary to the United States to meet with Cromwell, but shortly after he arrived in New York, the plot was leaked to the Colombian treaty negotiator. When Amador tried to meet with Cromwell, the old man was literally thrown out of Cromwell's office. In desperation, Amador turned to Bunau-Varilla, who had arrived in New York. Bunau-Varilla, sensing that his beloved Panama Canal was again in danger, was about to play yet another pivotal role in its convoluted history.

Though Bunau-Varilla had no official position, he met separately with President Roosevelt and Secretary of State John Hay and discussed what was happening in Panama. All the parties later denied that he had ever asked for or been promised help for the revolutionaries. But Bunau-Varilla left with the impression that the United States government would look favorably on a Panama revolution, an impression Roosevelt later confirmed. For their part, the revolutionaries, eager for United States help, were happy to use a canal treaty as a bargaining chip to secure it.

On November 2, the United States gunboat *Nashville* arrived in the harbor of Colón, on the Atlantic side of the Isthmus. On arrival, it found secret instructions to "maintain free and uninterrupted transit" on the railroad. If any force tried to interrupt transit, the *Nashville* was ordered to "occupy the line of the railroad, prevent landing of any armed force with hostile intent, either government or insurgent."

On November 3, the conspirators launched their revolution. There were many bizarre and colorful happenings during this almost comic-opera revolt, one of the most ingenious involving Colombian troops from the warship *Cartagena,* which arrived in Colón the same night as the *Nashville.*

General Juan Tobar of Colombia landed with five hundred troops to put down the rebellion. He was met by Colonel James Shaler, the superintendent of the railroad, who received Tobar and his aides with great cordiality. He apologized that most of the railroad cars were presently in Panama City, on the Pacific side of the Isthmus. (Knowing that Tobar would attempt to put down the uprising, railroad officials had moved most of the railroad equipment across the Isthmus two days before.) Shaler told the general that he had arranged for a special one-car train to take him and his officers immediately. His troops would be sent along on the next available train.

In one skillful move, the Colombian armed forces had been decapitated. In Panama City, the officers were met by an honor guard drawn from the local Colombian garrison. As the honor guard filed past the unsuspecting officers, they suddenly turned their rifles on the officers, taking them prisoner. The revolutionaries had bought the soldiers off before the officers arrived.

By November 6, the revolution was over. There were only two casualties, both inflicted by the Colombian gunboat *Bogota,* which had fired a half dozen shells into Panama City on the first day of the uprising before being chased away by return fire from the revolutionaries. One of the casualties was

a Chinese shopkeeper asleep in his bed. The second was a donkey in a slaughterhouse. American troops fired not a single shot.

The Republic of Panama extended from Costa Rica to the Colombian border, deep in the Darien jungle. The new country covered less than thirty thousand square miles, about the size of South Carolina. At independence, its population was about 350 thousand.

Before the revolution, Bunau-Varilla had insisted on being made the official envoy to Washington of the new Panamanian government, making it a condition of helping the revolutionaries. They had reluctantly agreed.

Fearing that the newly independent Republic of Panama might prove as intransigent as Colombia

△ *Because of silting and landslides, dredging at the Canal is a never-ending task.* ▷▷ *In Culebra Cut, later renamed Gaillard Cut, workers used Bucyrus steam shovels to heap eight tons of rock and dirt in a single scoop onto flatcars.*

had been, Bunau-Varilla hastily signed a treaty with Secretary of State Hay in Washington on November 18. He did so knowing that a Panamanian delegation had arrived in New York to negotiate the treaty, and that it had explicitly forbidden Bunau-Varilla from making a treaty without its consent.

The treaty's financial arrangements were the same as those that had been offered to Colombia: ten million dollars plus an annual payment of 250 thousand dollars, a large sum in those days. But Bunau-Varilla wanted to be sure the treaty would be approved by the United States Congress, so he made sure other provisions were more generous. For one

thing, the Canal Zone, the region of American control along the canal, was to be ten miles wide, not six. And American rights to the zone were to be granted "in perpetuity," instead of in a hundred-year renewable lease.

Colombia had never been required to give up sovereignty over the zone, and neither would Panama. But the Hay–Bunau-Varilla Treaty had a peculiar twist: The United States would be allowed to act *as if* it were sovereign, with the same "rights, power or authority" it wielded over its own lands. One might consider the Panamanians' guarantee of sovereignty under these conditions a distinction without a difference.

Two hours later, the Panama delegation arrived by train from New York. Bunau-Varilla met the

△ *The Emberá-Wounaan, two related indigenous peoples who live mostly in the Darien jungle, traditionally wore few clothes. Now it's T-shirts and jeans, but on special days they wear their finest—sometimes little more than a flower.*

group at the train station. When he told them the news, one of the delegates is said to have struck him across the face.

Nevertheless, in the following weeks, Bunau-Varilla continued to act behind the delegation's back, cabling Panama with the threat that the United States would withdraw its protection of Panama if the government did not ratify the treaty. This was not true, but fear of reprisal from Colombia drove the Panamanian government to approve the treaty on December 2, 1903. The United States Senate ratified the treaty on February 23, 1904, by a vote of sixty-six to fourteen.

The Builders

. . . I am going to make the dirt fly!
—President Theodore Roosevelt

On November 12, 1904, the first American steam shovel bit into the earth at Culebra. At first, the American effort seemed destined to end as catastrophically as the French one had. Cautioned by the French scandal, the Americans managed the work from Washington, burying the simplest supply request in red tape. Through scrupulous accounting and hard-nosed frugality, they were determined to ward off any accusations of corruption.

In addition, the chief engineer hired to oversee the work, John Findley Wallace, turned out to be unsuited for the job. He had no clear idea how he was going to build the canal, and he spent the first year experimenting with construction equipment.

He also brushed aside the pleas of his sanitary officer, Colonel William Crawford Gorgas, for men and money to control the mosquito population. This in spite of the fact that Gorgas was an Army doctor who was considered a leading authority, if not the leading authority, on tropical diseases; he was the one who had been responsible for wiping out yellow fever in Cuba, the first country in the world to rid itself of the disease.

Working under Walter Reed, in 1901 a skeptical Gorgas led a cleanup campaign to eradicate the mosquito *Stegomyia fasciata* (now known as *Aëdes aegypti*) from Havana. A painstaking process, it meant sealing off and fumigating the houses of the sick, draining even the smallest pool of standing water, and making sure every water container in the city was covered or screened off with wire netting.

By 1902, yellow fever had disappeared from Cuba. The campaign also seriously damaged the *Anopheles* mosquito population, rightly suspected of causing malaria. By the end of Gorgas's campaign, the number of malaria cases had been cut in half.

Now a believer, Gorgas wanted to conduct a similar cleanup at Panama, before the influx of workers with fresh, nonimmunized blood arrived to set off new epidemics of yellow fever and malaria on the Isthmus. Chief engineer Wallace dismissed his request for the resources to conduct this campaign, telling him it had nothing to do with building a canal.

The end of November saw the first case of yellow fever. In the following months a minor epidemic hit the Isthmus, stirring panic among the workers. Outbreaks of malaria, pneumonia, tuberculosis, smallpox, dysentery—even a couple of cases of bubonic plague—did not help matters. Many fled for home, including, it would appear, Wallace, though he denied that was why he left Panama with his wife in June 1905, to attend to "matters of importance" in New York. William Howard Taft, at that time secretary of war and the man with ultimate authority over the work in Panama, promptly fired him.

The new chief engineer was a handsome, mustachioed railroad engineer named John Stevens. The first thing he did when he arrived on the Isthmus was halt all work on the Canal. Instead, he directed thousands of workers to build houses, hospitals, sewage systems, schools, mess halls—entire towns sprang up. He gave Gorgas his unlimited support, and backed it up with men and material.

Gorgas undertook perhaps the most ambitious sanitation campaign in civilization's history. In a year and a half, he and his team of four thousand men eradicated yellow fever from the Isthmus of Panama. Several years later, he brought malaria under control.

Now the work could truly begin. On June 21, 1906, Congress decided that the Panama Canal should be a lock canal. The man who convinced Congress was Stevens, who until shortly before the vote had favored a sea-level canal. Then he saw the Chagres during the rainy season.

The plan, as worked out by Stevens and his successor, George Washington Goethals, was this:

Instead of a ditch through the jungle, the Canal would be a kind of water elevator linked by man-made lakes. Locks would take ships entering the Canal from one ocean and lift them eighty-five feet above sea level, releasing them to navigate through fifty miles of flooded mountainous jungle. At the opposite end of the Canal, the ships would be lowered back down to sea level, where they would sail off into the other ocean.

But even though the Canal did not have to be dug down to sea level, the amount of excavation required was mind-boggling. By the time the Canal was opened in 1914, the Americans had moved 232 million cubic yards of earth in addition to the 30 million cubic yards of usable excavation completed by the French. This was three times the volume of excavation completed at Suez, a sea-level canal one hundred miles long. To put it another way, if all the spoil dug at Panama were piled into railroad dirt cars, the resulting train would be long enough to circle the earth four times at the equator.

At the height of the work, nearly 50,000 men and women were employed by the Isthmian Canal Commission, of whom 5,000 were Americans. Only 357 Panamanians ever chose to work on the Canal during the entire construction era. Most workers, particularly unskilled laborers, came from the West Indies—Barbados in particular. Their work was among the hardest and most dangerous; their living conditions, unhealthy and far inferior to those of the Americans. Still, they were better paid and cared for

△ *The Republic of Panama comes to a virtual halt for the Carnaval celebration. When Panama gained its independence in 1903, 350 thousand people lived on the Isthmus. Today, it has a population of approximately three million.*

at the Canal than they had ever been at home, which is why they came in droves to the Isthmus.

The most difficult digging took place at Culebra, the nine-mile-thick section of the Continental Divide that had plagued the French. Even with a lock canal, this wall of rock, clay, and dirt had to be broken through.

The man in charge of the work at Culebra was Major David Du Bose Gaillard, after whom the Cut was eventually named. But the ingenious plan for excavating it originated with Stevens, the veteran railroad engineer. He devised an incredibly elaborate system of rail lines for transporting the spoil. About 160 trains ran in and out of the narrow construction area at one time, on seventy-six miles of track that was constantly being moved as the work progressed. The web of rail lines amounted to an intricate conveyor belt. As a giant Bucyrus steam shovel—up to sixty-eight were used at the Cut at once—dug out an eight-ton scoop of rock and dirt, an empty dirt car was ready to receive its load and whisk it away. Millions of cubic yards of spoil were used to fill in swamps or to build a three-mile breakwater at the Pacific end of the Canal.

As the Cut grew deeper, workers used a cranelike machine called a track shifter to pick up and move whole sections of track to a fresh location. A dozen workers could move a mile of track in one day using the machine, a job that would have required six hundred men doing the work by hand.

△ A monument to Gaillard once stood on the Canal banks. Cut widening forced its relocation to the Administration Building. ▷ Work crews in Culebra Cut (now Gaillard Cut) shift tracks to accommodate trains hauling spoil away.

In spite of the progress the Americans made at the Cut, they were plagued by landslides just as the French had been. Workers dug wider and wider slices from the mountains as the engineers searched for the "angle of repose," the point at which the earth would finally come to rest. In doing so, the distance across the top of the Cut grew from 670 to 1,800 feet, adding twenty-five million cubic yards to the total excavation needed. Landslides still occur today, though less frequently than during construction days. The angle of repose has never been found.

Besides the Cut, perhaps the most miraculous engineering feat at the Canal was the building of the

locks. Each lock chamber in the Panama Canal is 1,000 feet long, 110 feet wide, and 81 feet deep. The twelve chambers, made from concrete poured into massive forms, took four years to build.

When in use, the lock chambers are kept partly filled with water, so it is hard to fathom how gigantic they are. But if a single chamber were upended and set down in Manhattan today, it would be one of the tallest structures dominating the skyline.

Stevens did not see the work completed. In early 1907, he resigned, for personal reasons never fully explained. Roosevelt replaced him with Goethals, an outstanding Army engineer. Though the Canal remained a civilian government undertaking, having an Army man in charge appealed to Roosevelt after losing two chief engineers: from now on, the chief could not leave without the president's permission.

The bulk of the work was completed under Goethals, a demanding man who gradually won the respect of a labor force that had adored Stevens. He proved a brilliant replacement. There were still many daunting setbacks and hardships ahead, but the plan was so sound, the engineering so brilliant, and the builders so hardworking that when Goethals took over, the completion of the Canal was never in doubt. The Canal was officially opened on August 15, 1914, less than ten years after the first American workers arrived on the Isthmus.

The Land Divided—The World United

> The creation of a water passage across Panama was one of the supreme human achievements of all time, the culmination of a heroic dream of four hundred years and of more than twenty years of phenomenal effort and sacrifice. The fifty miles between the oceans were among the hardest ever won by human effort and ingenuity, and no statistics of tonnage or tolls can begin to convey the grandeur of what was accomplished. Primarily the canal is an expression of that old and noble desire to bridge the divide, to bring people together. It is a work of civilization.
> —DAVID MCCULLOUGH,
> The Path Between the Seas

The American effort cost $352 million, including the $10 million paid to Panama and the $40 million

paid to the French Company. It was by far the most expensive construction project in the history of the United States up to that time.

Building the American canal also cost 5,600 lives, most of these West Indian. Only 350 Americans died on the Isthmus during the ten-year construction period.

Though a tragic human toll, the total number who died was only a quarter of the number lost during the French effort. The Isthmus was a far healthier place to live under the Americans than it had been under the French, and the credit for this belongs squarely with Gorgas.

Throughout the more than eighty years since the Canal opened, improvements have continued to be made: Gaillard Cut, 300 feet wide when the Canal opened, was widened to 500 feet by 1971. Channel lighting was installed along the canal in 1966, allowing ships to transit twenty-four hours a day. By 2001, new excavation had widened the Cut to 630 feet at its narrowest point.

But in most respects, a Canal transit today is the same as it was in 1914. The Canal was so meticulously crafted, and has been so well maintained, that most of the original equipment is still used today, right down to the shiny aluminum handles mounted on blue marble slabs in the control houses, switches that open and close the lock gates with a turn of the wrist.

And the eternal jungle that greets the ships as they approach Limón Bay, the Atlantic entrance of the Canal, looks much the same as it did on Christmas Day 1502, when Columbus took shelter from a storm there. Columbus never found a way across the Isthmus, but today more than twelve thousand oceangoing ships do so each year.

The best way to grasp the way the Canal works is to imagine a typical transit. Cruising through Limón Bay, a ship is met—not by land—but by the two massive gate leaves, each sixty-five feet wide and seven feet thick, of the first "miter" or lock gates of Gatun Locks. Like the double doors of a giant's house, the leaves swing open silently in the middle and come to rest in recesses in the lock walls. The ship motors in, guided by lines attached to electric towing locomotives running in tracks along the sides of the chamber. The miter gates swing closed behind the ship. A worker in the

control house opens a valve that allows freshwater from the man-made Gatun Lake to roar down culverts, eighteen feet in diameter, which run below each lock chamber. The water flows into smaller cross culverts, finally coming up through one hundred holes at the bottom of the chamber, lifting the ship to the level of the next chamber. Filling the chamber takes about ten minutes.

The gate in front of the ship then swings open, and the ship moves forward into the second chamber, where the process is repeated, and then it moves to the third and final chamber. By the time the ship leaves the mile-long Gatun Locks, it has been lifted eighty-five feet above sea level. This herculean effort, as impressive today as it was in 1914, is accomplished solely by the force of gravity.

◁ A French bucket dredge works to clear a slide in the Gold Hill area of the Cut shortly after the Canal opens. Slides still occur. △ Time has stood still for many Emberá and Wounaan, who live in the Darien jungle near the Colombian border.

The ships are raised or lowered by the pressure of water flowing from higher to lower elevations. No pumps are used.

Each complete canal transit requires fifty-two million gallons of water from the lakes, which is lost to the sea. Fortunately, Panama's eight-month rainy season means that there is rarely a water shortage.

The ship then enters Gatun Lake, created by damming the Chagres River near its mouth on the Atlantic side of the Isthmus. Gatun Dam, an earthen structure one and a half miles long and a half mile wide at the base, was the world's largest earthen dam when it was completed, and it created the then-biggest manmade lake in the world, flooding 163 squares miles of jungle, construction-era towns, evacuated villages, and the hard-won former route of the Panama Railroad. Gatun Lake finally solved the age-old problem of the Chagres by submerging that part of it which crossed paths with the Canal. No matter how high the Chagres gets in the rainy season, the level of the lake is controlled by releasing water through a spillway in the dam.

Our ship sails about twenty-three miles across Gatun Lake, following the riverbed of the Chagres, before entering Gaillard Cut, the nine-mile-long gap in the Continental Divide, where mountains loom above the ship.

△ Ship captains must turn control of their vessels over to pilots of the Panama Canal Commission during a transit of the Canal. Depending on the size of the ship, up to four pilots may be needed for each transit.

Next, the ship enters Pedro Miguel Locks, where it is lowered thirty-one feet to the level of Miraflores Lake, a second man-made lake.

At the opposite end of the mile-wide lake is the last set of locks, Miraflores. The ship is lowered fifty-four feet in two lockages, back down to sea level. It then sails out past the headquarters of the Panama Canal Commission at Balboa, under the soaring Bridge of the Americas that spans the Canal, past Panama City, and out into the Pacific Ocean.

The whole transit, the realization of a dream more than four hundred years old, takes only eight to ten hours.

The Future

This gringo is going home.
  —Popular T-shirt after the ratification of the
    1977 Carter-Torrijos Treaty

For many decades, the Canal operated quietly and efficiently, like the well-oiled machine it is. And a self-sufficient one at that: by United States law, the Canal supported itself through tolls paid by transiting ships. Ships have made approximately nine hundred thousand transits of the Panama Canal since 1914, and now move nearly two hundred million tons of cargo over the mountains of Panama each year.

Over time, a unique Zonian way of life evolved near the banks of the Canal. For the most part, it was a simple, orderly, placid life. Nothing much ever changed, and it seemed as though it never would.

Meanwhile, Panama was growing up. In part because of the presence of the Canal and the fact that it has the U.S. dollar as its currency, it developed into one of the most prosperous countries in Latin America. It became known as the "Switzerland of Central America," a financial center that at one time boasted more than one hundred international banks.

And right down the middle of this growing country ran a foreign community fifty miles long and ten miles wide. In retrospect, it could never have lasted.

And it did not. The decision to abolish the Canal Zone and transfer control of the Canal to Panama is a long, complex, and controversial one, a saga in itself. But on September 7, 1977, Jimmy Carter, president of the United States, and General Omar Torrijos Herrera, dictator of Panama, signed a treaty that began the transition of the Canal to Panamanian control.

The Canal Zone disappeared from the world's maps at midnight on October 1, 1979. The treaty called for the Canal itself to be turned over to the Republic of Panama in stages, during a twenty-year transition to end at noon on December 31, 1999. This slow relinquishing of control has been called "the long good-bye."

In the interim, the treaty established a new U.S. government agency, the Panama Canal Commission, to run the Canal. Americans were gradually phased out of the workforce of the Canal in favor of Panamanians during the transition years. Today, out of a Canal workforce of nine thousand, less than one

28

percent is American. Most American families that lived and worked in the Canal Zone for many years, sometimes for generations, said their own long, sad good-byes and left to make their homes elsewhere.

Panama saw more than its share of turmoil during those transition years. It endured both the repressive military dictatorship of General Manuel Antonio Noriega and the U.S. invasion, dubbed "Operation Just Cause," that violently removed him from power in 1989. Throughout this dark period, however, the Canal was allowed to operate without interference.

The turn of the new century began a new era for the Panama Canal. A generation of Panamanians has come of age since the signing of the 1977 Panama Canal Treaty, a generation determined to show it can run the Canal as well as the Americans did.

The Canal was essentially the same on January 1, 2000, as it had been on December 31, 1999. The main differences were that the organization that ran it had a new name, La Autoridad del Canal de Panamá (ACP), or Panama Canal Authority, and answered to the Panamanian government, not the U.S. But the ACP is an apolitical and largely autonomous entity. In fact, the administrators of the ACP are consciously shaping it to more closely resemble a business than a government agency. Being a nonprofit service to the world is no longer the goal; making money is. In July 2003 the ACP phased in a 12.5-percent hike in tolls, only the ninth time in the Canal's history that tolls have been raised.

It will probably be years before the shape of the Canal's future becomes clear. But change is coming fast to the once-immutable Canal and former Canal Zone. Panama has inherited a huge chunk of real estate, and it is developing parts of it rapidly. The jungle-draped townsite of Gamboa, home to the Canal's Dredging Division, has been reborn as an ecotourism resort. A Hong Kong company has revamped the old ports of Balboa and Cristobal to handle huge stacks of cargo containers. New ports have been constructed. The historic Panama railroad, allowed to rust away under Panama's military dictatorship, has been rebuilt by an American company. Malls and mansions are springing up at what once was Albrook Air Force Base. Fort Clayton, a former U.S. Army base, has been reborn as something called the City of Knowledge, which aims to be an international center for education and research.

At Fort Amador, another former Army base at the Pacific entrance to the Canal, developers are building hotels, restaurants, condos, shops, a cruise-ship port, and a marina and have plans to build a futuristic museum designed by Frank Gehry. Most of this is being built along the Causeway, a breakwater that once boasted little more than a community theater, a Smithsonian research lab, a small beach, and a lover's lane for high-school students.

Panama City is engulfing much of the former Canal Zone, with multilane freeways arcing toward its quiet, tree-lined streets. A second bridge spanning the Canal is being built at Pedro Miguel to relieve the enormous strain traffic is putting on the Bridge of the Americas, near the Canal's Pacific entrance. Sadly, Balboa—the de facto "capital" of the Canal Zone—

△ *Panama City is a fast-growing metropolis, with new hotels, condominiums, and office buildings seeming to spring up overnight. The Canal, beaches, casinos, and year-round temperatures above 80°F make it a tourist destination.*

is turning into an industrial ghetto, hemmed in by highways, train yards, and giant stacks of ship containers. Other parts of the Zone have become ghost towns, with the jungle that was held back at such enormous cost of lives and money again winning the battle. In a few years, one-time Zone residents may find their hometowns no longer exist.

Changes at the waterway itself are even more dramatic. The Canal is preparing aggressively for the demands of twenty-first-century commerce. It has started with a billion-dollar modernization program that has transformed the look and operation of much of the Canal.

Central to this program was the widening of Gaillard Cut, which was completed at the end of 2001. Dredges and dump trucks carved away at the Cut, widening it from 500 feet to a minimum of 630 feet, 730 feet on curves.

The Cut was widened because the percentage of large ships transiting the Canal is increasing. Large ships have traditionally not been allowed to pass each other in Gaillard Cut for safety reasons. Cut widening, a controversial project that critics say has still not proved its worth, was meant to permit even the biggest vessels allowed in the Canal, "PANAMAX"–size ships, to slip past each other.

The locks are also being transformed, though in a less noticeable way. The ingenious machinery that opened and closed gates and valves for nearly a

△ *The Ngöbe-Buglé, or Guaymi, are the most numerous of Panama's indigenous peoples. Girls and women still wear traditional dresses in their everyday lives.* ▷ *Teddy Roosevelt worked on a ninety-five-ton steam shovel on a visit in 1906.*

century has been replaced by hydraulics and computer controls.

In 2002 the Canal began an ambitious project to deepen, by three feet, the navigational channel that runs down the middle of Gatun Lake. This project will also increase the capacity of the Canal's water reservoir by 25 percent. The project is estimated to cost $190 million.

Other changes include a new vessel traffic-management system that uses satellite Global Positioning System technology and a new fleet of towing locomotives that will run on fifty thousand feet of new tow track.

Even before all these changes, the Canal was able to accommodate more than 90 percent of the world's ocean-going vessels at the end of the twentieth century. And the Canal is far from becoming obsolete. It's a crucial part of Panama's economy, which is dominated by a services sector that also includes one of the world's largest free zones, an international banking center, maritime services, and a growing tourism industry. In fiscal year 2002, even with the downturn in the world economy, the Canal set a new record for total revenues: $589 million, of which $271 million was handed over directly to the central Panamanian government.

Today the Canal faces the daunting task of projecting the needs of twenty-first-century world commerce. It is studying the feasibility of building a new, larger set of locks to parallel the first two sets. Such a project was actually begun under the Americans but abandoned when the U.S. entered World War II.

Doubts remain, however, about the necessity or value of the project. Some argue that the potential tolls raised could never justify the enormous cost. Then there are the logistical concerns: For one thing, a third set of locks would mean greatly expanding the Canal's watershed. The Canal is already having trouble preserving the current watershed, which is threatened by deforestation, encroaching urbanization, and pollution.

There are some who think even bigger, again floating the age-old dream of a sea-level canal. If this ever does come to pass, it would certainly be one of the great achievements of this century. And chances are good it, too, would be built in Panama.

Those who contemplate the second hundred years of the Panama Canal can't help but find parallels with its first hundred years: the incredible challenges, the great risks, the potential triumphs.

But all these now belong to the Republic of Panama, whose success in operating the Canal will be vital to its prosperity as a nation.

Though we no longer play a part in the future of the Panama Canal, many of us who once made our homes near its banks find that every change stirs up bittersweet memories. We discover we still care deeply for a waterway and a little world we once took for granted. For us, the long good-bye continues.

◁ "I come not hither to hear lamentations and cries, but to seek Moneys." Welsh pirate Henry Morgan sacked the old Spanish city of Panama in 1671. Stone ruins still stand a few miles from the highrises of modern-day Panama City. △ The Administration Building, on a hill overlooking the former Canal Zone, is the headquarters of the agency that runs the Canal. A growing Panama City is gradually encroaching on the Canal area.

△ Fresh water thunders through the spillway of Madden Dam, which together with Gatun Dam regulates the level of Gatun Lake and generates hydroelectric power for the Canal and its environs. ▷ A tug acts as an auxiliary rudder for a car carrier negotiating the curves of Gaillard Cut. ▷▷ Electric towing locomotives guide ships through the locks, running along tracks on both sides of the lock chambers.

△ The widening of Gaillard Cut, completed in 2001, was part
of a billion-dollar modernization program. The drillboat *Thor*
was used to drill holes for underwater blasting.

△ As a ship approaches a set of locks, Panama Canal Commission linehandlers on land, ship, and traditional Panamanian rowboats called *pangas* still secure the cables to ships and locomotives in the same way they have since the Canal opened in 1914: by hand. ▷▷ Hurling heavy lines between ship and shore is a dangerous, precise job.

◁ In the west lane of Gatun Locks, Panama Canal employees perform maintenance work in a dewatered lock chamber while bulk carriers transit in the east lane. △ A busy evening at Gatun Locks is a reminder that work at the Canal never stops. ▷▷ Planned preventive maintenance is the key to the Canal's efficient operation.

△ A computerized ship-handling simulator helps train pilots and tugboat masters to confront the unique challenges of navigating the Canal. Besides maneuvering through the locks and Gaillard Cut, a transiting ship must follow the route of the submerged Chagres riverbed, hidden beneath Gatun Lake. ▷ Ships transit the Canal twenty-four hours a day. A southbound container ship approaches the upper chamber of Miraflores Locks.

◁ The Panama Canal is one of the world's most popular cruising destinations. Transits last a full day. △ A cruise ship glides across Gatun Lake, a freshwater lake eighty-five feet above sea level. It was once the largest man-made lake in the world.

△ A bulk carrier justifies its name at Miraflores Locks on a southbound transit. Ships with a beam (width) of up to 108 feet are allowed in the locks. The lock chambers are 110 feet wide. ▷ A sailboat follows the car carrier *Sunbelt Spirit* into Gatun Locks. The closed-lock gate allows a vehicle to scoot across the retractable bridge. ▷▷ At times, tugboats accompany smaller ships into the locks, acting as a kind of outboard rudder.

△ Crossing the top of the miter gates can be a harrowing experience for the uninitiated, but it's just another day at work for employees at the locks. ▷ The massive lock gates of the Panama Canal weigh up to 745 tons each but were so ingeniously designed that a forty-horsepower motor could open and close them. ▷▷ Workers overhaul lock chambers one lane at a time, allowing transits to continue in the other lane.

△ Ships are raised and lowered eighty-five feet in three lockages at Gatun Locks, the mile-long set of locks near the Canal's Atlantic entrance. A bird's-eye view shows the difference in elevation between Gatun Lake and the Atlantic. ▷ A steel ribbon linking the continents, the mile-long Bridge of the Americas soars nearly four hundred feet above the Canal's Pacific entrance. ▷▷ Over 150 thousand cruise passengers transit the Canal each year.

△ Those who have never visited the Canal tend to picture it as a great ditch. Only Gaillard Cut, not the locks, might fit that description. ▷ A better analogy would be the world's largest water elevator, lifting ships eighty-five feet above sea level at one end and lowering them back down to the sea at the other.

△ Even the mammoth culverts, eighteen feet in diameter, that run beneath the lock chambers must be maintained. During the transit, the turn of a handle sends fresh water roaring down the culverts to fill or empty the lock chambers. ▷ Water flows through one hundred holes at the bottom of each chamber, ensuring that ships are raised and lowered smoothly.

◁ A roll-on/roll-off cargo ship squeezes through Pedro Miguel Locks on a northbound transit. △ Ships have made about 900,000 transits across the Isthmus of Panama since the Canal opened in 1914. ▷▷ For nearly one hundred years, Kuna women have decorated their blouses with *molas,* cotton panels they make by cutting and stitching layers of brightly colored cloth to depict primitive figures and elaborate abstract patterns.

△ A memorial to chief engineer Colonel George W. Goethals stands below the Administration Building in Balboa Heights. ▷ Panama is a sun-lovers' paradise, with hundreds of miles of nearly deserted sandy beaches and countless tropical islands.

▷▷ The widening of Gaillard Cut was the most ambitious construction project at the Panama Canal since it opened in 1914. Entire hillsides were leveled. The Cut was widened from 500 feet to a minimum of 630 feet. To allow large ships to maneuver around curves, in some places it is now 730 feet wide.

△ Artwork on wheels, the brightly painted buses of Panama often reveal the interests of the proud owner/driver. Many include portraits of the latest American movie star or popular musician. ▷ A folkloric dancer at the ruins of Old Panama wears a *pollera,* the gorgeous, hand-embroidered national dress of Panama. ▷▷ For the sake of economy, yachts and other small vessels transit the Canal in groups.

◁ The Kunas, like the Emberá and Wounaan of the Darien jungle, hold on tightly to many of their traditions. Panama's other indigenous peoples include the Ngöbe-Buglé (aka Guaymi) and the Naso (aka Teribe). △ Christopher Columbus took shelter from a storm in Limón Bay on Christmas Day, 1502.

Panama Canal Facts

Average time spent transiting the Canal: 8 to 10 hours.

Fastest transit time, set by the U.S. Navy hydrofoil
Pegasus: 2 hours, 41 minutes.

Average Canal toll paid by oceangoing
commercial vessels: approximately $50,000.

Record toll as of January 17, 2003, paid by the cruise ship
Coral Princess: $217,513.75.

Lowest toll ever paid, by the adventurer Richard Halliburton,
who swam the Canal in 1928: 36 cents.

Number of other adventurers allowed to swim the Canal today: 0.

Width of each lock chamber: 110 feet.

Beam of USS *New Jersey* and her sister ships,
the widest allowed to transit: 108 feet.

Height of Empire State Building: 1,250 feet;
length of a lock chamber: 1,000 feet.

△ *The lock chambers of the Canal can still accommodate 92 percent of the world's ships. The lock walls were made from concrete poured into gigantic steel forms. The lock gates were made in Pittsburgh.*

Length of the bulk/oil carrier *San Juan Prospector,*
the longest ship to transit: 973 feet.

Percentage of the world's ships too large to transit the Canal: 8.

Approximate number of transits that have been made
through the Canal: 900,000.

Amount of work on the French canal actually done
by Americans: one-third.

Number of people killed during Panama's 1903 revolution: 1.

Number of countries represented by the Canal-building
labor force: 97.

Number of Panamanians on that force: 357.

Number of Barbadians: 20,000.

Major user of the Canal: United States.

Most common cargo to transit the Canal: grain.

Most common trade route of transiting ships:
between the eastern United States and the Far East.

Approximate percentage of transiting ships taking this route: 43.

Depth of the lock chambers: 81 feet.

Maximum ship draft allowed in the locks: 39.5 feet.

Number of water pumps used to raise and lower ships: 0.

Number of gallons of water needed to transit a ship:
52 million.

Diameter of main water culverts beneath the lock chambers:
18 feet.

Diameter of the Hudson River tubes of the Penn Central
Railroad: 18 feet.

Number of cubic yards of concrete needed to build the
Canal's locks and dams: 4.5 million.

Savings when builders were ordered to shake concrete sacks
after emptying them: $50,000.

Most transits on a single day, set on February 29, 1968: 65.

Approximate difference in the "heights" of the
Atlantic and Pacific Oceans: 0.

Approximate tide size on the Pacific side of the Isthmus:
18 feet.

Approximate tide size on the Atlantic side of the Isthmus:
18 inches.

Number of miles a ship traveling between New York and
San Francisco saves by using the Panama Canal
instead of going around Cape Horn: 7,872.

Distance *east* traveled by a transiting ship going from the
Atlantic to the Pacific Ocean: 22.5 miles.

Number of times a train filled with dirt and rock dug from
the Canal would circle the earth at the equator: 4.

Sources and Suggestions for Further Reading

Howarth, David. *The Golden Isthmus.* London: Readers
Union/Collins, 1967.

Knapp, Herbert and Mary. *Red, White and Blue Paradise:
The American Canal Zone in Panama.* New York: Harcourt
Brace Jovanovich, 1984.

McCullough, David. *Path Between the Seas: The Creation of
the Panama Canal 1870–1914.* New York: Simon and
Schuster, 1977.

The Panama Canal: A Vision for the Future. Balboa, Republic
of Panama: Panama Canal Commission Printing Office,
1997.

Panama Canal Review and other publications of the Panama
Canal Company and Panama Canal Commission,
1969–1998.